BORN
Wild

in YELLOWSTONE
and GRAND TETON
NATIONAL PARKS

Henry H. Holdsworth

FARCOUNTRY
PRESS

This book is dedicated to
my wife Valerie and
my daughter Avery.

Right: A cow and calf moose splash across the shallows of the Snake River on a beautiful fall morning at Oxbow Bend.

Title page: Tree huggers! Twin black bear cubs cling to the safety of a lodgepole pine tree as their mother grazes on plants close by.

Front cover: First kiss: A cow and calf moose share a tender moment while resting in the tall grass. The mother's affection for her new-born is evident, and she will do her best to protect it from predators such as grizzlies, coyotes, and wolves.

Back cover: Black bear cub (left) and a family of curious baby Uinta ground squirrels (right).

ISBN: 1-56037-245-1
Photographs © Henry Holdsworth, and others as noted
© 2003 Farcountry Press

For more information on our books write: Farcountry Press, P.O. Box 5630, Helena, MT 59604 or call: (800) 821-3874 or visit www.montanamagazine.com

Created, produced, and designed in the United States.
Printed in Korea.

Foreword

by Henry H. Holdsworth

A young pika bathes in the sunlight of a late summer morning. These cute little mammals can be found on rocky slopes at higher elevations.

Ah! Springtime in the Rockies. There is no season that is more anticipated for those who live in the mountains. It is a time when winter eases its icy grip on the land, a time of renewal, a time of rebirth and awakening. It is a time when color returns to a landscape that has been blanketed in white, when the voices of songbirds can be heard again and the scent of sage is in the air. Spring is a time of lengthening days, warm breezes, and cool nights, when long dormant signs of life begin to show themselves once more.

Wildflowers bloom, slowly moving higher as snow lines retreat skyward, and the buzzing of insects returns to once quiet airways. The clearing of ice from frozen lakes and ponds makes way for ducklings and gives us hope that summer is not far behind. These signs of spring are familiar to any mountain community, but in Yellowstone and Grand Teton National Parks it is the diversity of wildlife that helps set them apart.

Set amid lofty peaks and boiling geysers, Yellowstone and the Teton Range are at the heart of an ecosystem that is wild and intact. This is a place where grizzly bears still wander, mountain lions and wolves find safe haven, and great herds still roam. Each spring the baby animals of the parks take center stage as a new cycle of life begins and the magic of nature unfolds anew. From the diminutive ground squirrel to the magnificent moose, all creatures great and small, hoofed and winged, burst onto the scene and set the stage for one of nature's greatest shows. For in Yellowstone Country you can still witness meadows filled with orange baby bison and spotted elk calves, or flotillas of trumpeter swans and fox pups frolicking in the high grass. But for each tender moment there is also danger lurking at every turn, which makes this time of year one of unforgettable drama.

Spring comes later and more slowly to Yellowstone and the Grand Tetons than it does in most places. With valleys at 6,000 to 8,000 feet in elevation and peaks that range close to 14,000 feet high, the thought of snow is never far off and the change in season is often hard to define. The first hints of spring come in April, when mother bears emerge from dens with cubs in tow, bald eagle chicks break out of their eggs, and the first baby bison are born into a snowy world. By mid-May Canada geese goslings grace the park waterways, and by early June elk and moose calves are a common sight. As July approaches mule deer fawns might be seen chasing through mountain meadows, and baby birds of all shapes and sizes chirp for food at the nest. "Born Wild" takes you on a journey into the private lives of the parks' precious babies and captures moments from their first days of life through their first year. If you have never seen a grizzly cub play tag with a raven or a pronghorn's first steps, then the magic of Yellowstone awaits.

The pages of this book will shed some light on what it is like to be born wild in Yellowstone and Grand Teton National Parks.

Bison calves go nose to nose during a June snowstorm in Yellowstone National Park.

A buckrail fence makes a good place for this red fox kit to test out its agility. Young foxes have to practice hard in order to keep their reputation for being sly. As this pup gains more confidence in his abilities, he will venture greater distances from the den.

Above: The warm glow of reflected evening light surrounds this family of trumpeter swans. The small ponds and lakes of the Yellowstone Region make excellent nesting sites for these rare and stately birds.

Right: A great gray owl chick blends in with his surroundings. At two months this chick is not developed enough to fly but old enough to venture out of the nest.

Above: Killdeer nest and eggs during a late spring snowstorm. Hatching young in the high country is a challenge for all birds.

Right: This bighorn sheep lamb will spend most of its first few weeks of life in steep and rocky terrain to help keep it safe from predators. Bighorn are amazingly sure-footed.

Above: Twin grizzly cubs play pattycake along the banks of Soda Butte Creek in Yellowstone's Lamar Valley.

Facing page: An exuberant baby badger tries to coax his mother into a wrestling match. Its large claws will make it an excellent digger as it grows older. (DIANA STRATTON PHOTO)

Above: A yellow-headed blackbird chick chirps for food from the end of a cattail on the National Elk Refuge.

Left: New generations of bison and lodgepole pine appear together under the snow-capped peaks of the Gallatin Range. Fire is just one of the natural forces that are constantly at work in an ever-changing environment.

Above: Early morning light bathes a cow bison and her calf as they ford the Yellowstone River in Hayden Valley. River crossings are one of the hazards young animals must learn to master.

Facing page: For the first few weeks of life, this pronghorn fawn will spend most of its time lying low to hide from predators such as coyotes. It is truly amazing how well they can conceal themselves on ground that is mostly sparse grasses or sagebrush.

Above: A playful young marten peaks out through a hole in a lodgepole pine. His sleek body and semi-retractable claws make him well suited for life in the trees. He will spend most of his time hunting for squirrels, birds, and rodents in dense evergreen forests.

(DAN & CINDY HARTMAN PHOTO)

Right: A female red-tailed hawk watches over her young chicks. The red-tailed is one of the most common hawks in the region, and a healthy population helps keep rodent numbers in check.

(DONALD M. JONES PHOTO)

Two moose calves lounge in a meadow near Yellowstone Lake. Twins are a common occurrence in moose, the largest members of the deer family.

Jumping for joy, this black bear cub leaps at the opportunity to climb a tree.
Although both black bears and grizzly cubs can climb trees, adult black bears
are far better climbers than their grizzly counterparts.

It's bath time for this one-day-old bison calf. Even baby bison have to be reminded to clean behind their ears.

A cow elk nuzzles her newborn calf. This gentle washing will help keep the newborn free of scent and aid it in hiding from predators.

A female sandhill crane takes her two-week-old colt for a stroll through the dandelions.
Sandhills normally hatch out two eggs each spring, but usually only one chick survives.

Right: Two baby northern pygmy owls venture out from the safety of their aspen tree home. These tiny owls will be less than seven inches tall when fully grown.

Below: A baby beaver cruises the waters of the Snake River in Grand Teton National Park. Born in early spring, it is usually late summer before they surface from the lodge to feed and take short swims.

A moose and calf wander through the willows, their primary winter food source.
Moose have extra long legs to help them get through the deep snow of a Wyoming winter.

Above: Peekaboo bison. This calf dares a close-up view of the photographer behind the safety of its very large mother.

Facing page: Coyote pups await the arrival of their mother at the mouth of the den. Coyotes may have from three to nine puppies in a litter, and it is usually six to ten weeks before they emerge from the den for the first time. (DONALD M. JONES PHOTO)

Above: These baby saw-whet owls got off to a rough start in life. The tree they were nesting in, just outside of Grand Teton National Park, came down and their mother abandoned them. They were rescued and cared for until they could be released back to the wild in the fall.

Facing page: These spring cubs, or cubs of the year, will stay close to mom during their first few months. It is a dangerous world out there, even for baby grizzlies, which have to keep a watchful eye for wolves, coyotes, and male grizzlies. (DIANA STRATTON PHOTO)

Two cow elk escort a calf across the travertine terraces of Mammoth Hot Springs. The young animals of Yellowstone have to learn to maneuver themselves around the park's many interesting thermal features.

Whitetails are a rare sight in either park. This fawn and two does spent the summer in the Tetons and now winter along Fish Creek at the southern end of the range.

Canada geese escort their new goslings through the sparkling waters
of Christian Pond in Grand Teton National Park.

A bighorn sheep ewe and lamb take a stroll on a rocky outcrop. (DONALD M. JONES PHOTO)

Right: A family of badgers surveys the scene in Yellowstone's Lamar Valley. These youngsters will spend much of their time in the den while their mother makes constant hunting trips in search of ground squirrels, their main diet.

Below: All dressed up with no place to go! These baby American coot chicks show off wild colors compared to the drab-looking black adults that they will grow up to be.

Twin mule deer fawns skip with their doe through a meadow
in the Canyon area of Yellowstone Park.

Hang ten. A black
bear cub negoti-
ates its way down
from a nap in this
lichen-covered
evergreen tree.

Born in mid-April, eagle chicks are one of the earliest signs that spring is approaching. This two-week-old youngster will see more than one snow-storm in his first few months of life.

Frosted elk calf on a morning well below zero in Yellowstone's Upper Geyser Basin. This elk calf is well insulated from the brutal cold of a Yellowstone winter.

Lean on me! A grizzly cub gets a leg up from its sibling to get a better look at its surroundings. It's hard to see when you're this short. Cubs constantly stand on their hind feet to improve their view. Twins are common among bears, and one, two, three or, rarely, even four cubs can be born at once. (DONALD M. JONES PHOTO).

Left: A tiny killdeer chick pauses for a moment at the edge of a pond. The cry of the killdeer is a common sound of spring along the shores of Yellowstone's streams, rivers, and lakes.

Below: A baby yellow-bellied marmot sits patiently on a lichen-covered rock. Also known as "rock chuck" or "whistle pig" for its loud sharp call, this large member of the rodent family summers on rocky slopes and hibernates in winter. (DIANA STRATTON PHOTO).

Born with spots to help camouflage it from predators, this newborn mule deer fawn will spend most of its first few months hiding among downed timber deep in the forest.

By using the hot springs for warmth, this nine-month-old bison calf has already learned one valuable secret to surviving a frigid Yellowstone winter.

Chocolate moose. Moose calves are generally much lighter in color than their mothers. As the summer wears on, their fuzzy brown fur will gradually give way to the longer, darker guard hair that will get them through the winter.

On the run. A pair of elk calves put it into overdrive after lagging behind the rest of the herd. At this age they don't have to worry about bears and coyotes as much, but wolves are always a concern.

Left: Morning light illuminates the fuzzy fur of a baby Uinta ground squirrel. The park's most commonly seen mammal during the summer, they can be found in sagebrush habitat from mid-April through mid-August before they go into hibernation.

Below: Having your ducks in a row is the order of the day whenever this family of mallards sets out for a voyage.

Facing page: Stopping to smell the flowers, or dandelions as the case may be, a black bear cub takes a break. Baby bears and wildflowers are two of the highlights of any summer trip to Yellowstone. (DONALD M. JONES PHOTO).

A Canada goose checks her brood of freshly hatched chicks on the National Elk Refuge. Beginning in early May, goslings are a common sight on the region's waterways.

This bison cow, like many new mothers, struggles to keep her eyes open. Each time she dozed off, her massive head fell on her poor sleeping calf. After this scenario played out several times, the wise calf moved out of the way to sleep.

Feeling right at home on its winter range, this six-month-old bighorn calls out to its ewe.

Toeing the line. These grizzly twins stay in step with their mother
while causing a major "bear jam" in the Lamar Valley.

Right: A young mountain bluebird gets a little stretching in before taking to the air.

Below: A pair of trumpeter swans keeps a watchful eye over their newly hatched cygnet. The largest of all the world's waterfowl, trumpeters mate for life, and return each year to the same nesting sites to raise their young. They will generally lay between two and seven eggs in a clutch.

Facing page: It's always nice to be appreciated. This cow elk gets a late Mother's Day kiss from her two-day-old calf.

Yikes! That was a close call. A bison calf gets out of the way just in time, as two female bison settle a disagreement.

Left: A young beaver dines on the succulent bark of a cottonwood branch. Born into the largest member of the rodent family, he could weigh as much as 60 pounds when fully grown.

Below: Born to run. A newborn pronghorn tries out legs that seem a little wobbly. These same legs will one day take it to speeds up to seventy miles per hour, making it the fastest mammal in North America.

Mouth to mouth. One fox kit gains the upper hand in a wrestling match with its littermate. Play fighting is a valuable tool to these pups as they gain the strength and coordination they will need to survive as adults. (DONALD M. JONES PHOTO)

There is nothing like a quick dip in the pond on a warm summer day. Although this moose calf still gets most of its nourishment from its mother's milk, it will also take a few bites of aquatic vegetation if the mood strikes.

Above: A fleet of Barrow's goldeneye chicks takes a cruise on a channel of the Snake River in Grand Teton National Park. Goldeneyes are one of the few ducks that will remain year round in the park.

Right: A young river otter surveys the scene on the shoreline of Yellowstone Lake. A master fisherman, the otter is among the most fun loving and playful animals found in nature.

Like most of us around the Yellowstone Ecosystem, this bison calf is probably wondering if winter will ever end. Now ten months old in late February, he has made it through the toughest part of the winter and nearly all of the way through his first and most difficult year of life.

A young calf shows that even a moose can be pretty darn cute.

Right: This little golden ball of fluff will grow up to be a Canada goose.

Below: Twin elk calves frolic in Gibbon River meadows at the center of Yellowstone National Park.

Kissing cousins, or sisters, as the case may be, two baby Uinta ground squirrels enjoy each other's company just above the safety of their hole.

Looking as fuzzy as a stuffed animal, this bighorn sheep lamb is just about two weeks old.
(DONALD M. JONES PHOTO)

Timing is everything to the cycle of life in the high country. You don't want to be born too soon and get caught in a late spring blizzard, but you have to be born early enough to survive your first winter. Pictured here during an October snowstorm, this moose calf grew tremendously during its first five months. Reaching full maturity at about six years of age, it could one day weigh up to 1,200 pounds.

Just up from their long winter's nap, a grizzly sow and her cubs search for
food in a May snowstorm. These two-year-olds must now hope that their

A cow elk lovingly grooms her calf. They will be inseparable until the following May or early June when she gives birth again.

Left: A young least chipmunk dines in Grand Teton National Park.

Below: Catch me if you can. Red fox kits enjoy a high-speed game of tag.

Bear hug. These two-year-old grizzly cubs relish a little sibling rivalry in the form of a wrestling match. Bear cubs love to play, which is one reason they are so much fun to watch.

A baby porcupine finds safety up a tree. Porcupines are actually born with quills and their mothers must nurture them very carefully.

Mountain lion kittens roughhouse while their mother looks over her hunting grounds on the National Elk Refuge. Seldom seen due to their shy and nocturnal nature, these young cougars will become a valuable predator in the area's rugged and mountainous terrain.

Left: A common loon chick hitches a ride on the back of its parent. The remote backcountry ponds and lakes of the Yellowstone Ecosystem make excellent nesting habitat for the shy loon. Its call is one of the truly wonderful sounds of summer. (DONALD M. JONES PHOTO).

Below: A young moose calf hides in the sagebrush near Oxbow Bend in Grand Teton National Park. The shores of the Snake River and the Yellowstone Lake area are excellent places to find moose calves in June and early July.

A parade of elk calves cross the Gibbon River in Yellowstone Park. The explo-
sion of life in the high country makes spring a magical time to visit the parks.

Right: This tree swallow chick receives its breakfast on the fly.

Below: October flurries signal the arrival of winter for a six-month-old elk calf and its mother.

Practice makes perfect. Baby bison love to try out their moves in a game of play sparring. May is a great month to witness these little orange bundles of energy as they buck and scamper around the meadows of Yellowstone Park.

A brown-colored black bear cub peeks out from behind a rock. Black bears can also have cinnamon or black fur.

A moose calf finds a bite of grass to his liking. Mother moose are very protective of their young, so it is best to keep a safe distance when viewing a cow and calf.

HENRY H. HOLDSWORTH has spent over twenty years photographing and living in the Yellowstone Ecosystem. Educated as a biologist with a background in animal behavior, Henry has traveled extensively but there is no place he enjoys photographing more than his own backyard, Yellowstone and Grand Teton national parks. The birth of his daughter helped spark a renewed interest in the baby animals of the area and was the inspiration behind this book. *Born Wild* is Henry's fourth book on the region and is preceded by *Grand Teton National Park Wild and Beautiful* (2000), *Yellowstone & Grand Teton Wildlife Portfolio* (2001), and *Grand Teton Impressions* (2002). His baby animal pictures are also seen regularly in publications such as *Ranger Rick, Your Big Backyard, National Wildlife, National Geographic World, Wildlife Conservation,* and *Wyoming Wildlife.* Henry now divides his time between photographing and running his Wild by Nature Gallery in Jackson Hole. When he is not off capturing images, he can usually be found chasing his little girl Avery and his golden retriever Denali around his hometown of Wilson, Wyoming.